Starting from Your Head

Mental Geometry

by David Fielker

ISBN 1 874099 16 2

Typeset and designed by Fran Mosley
Cover design by Simon Spain, Orangebox Editions
Printed by GPS Ltd, Watford

CONTENTS

WHY MENTAL GEOMETRY?

Mental imagery

The subject of mental imagery has received some discussion in professional journals, but it gets scant attention in school schemes. This may be because it is difficult to build into traditional texts.

It is also because of a long-standing tradition of 'discovery' and 'practical' methods in which children are encouraged to *do* before they *discuss*, the doing being something involving physical action with materials. This is a sound principle, but it means that children are not often given the opportunity, or even encouraged, to think about things *before* they do anything with their hands, to *predict* what will happen and, in the case of spatial work, to *imagine* before they actually see.

This important ability to imagine, in the sense of 'to make an image of', is something that seems to decline with age if it is not practised. And the practice is generally limited to the few professions that make use of the power of imagery, like those of architect, sculptor and engineer. Three-dimensional work is particularly neglected, probably because it has long suffered from the two-dimensional restrictions of textual material that have tended to pervade all classrooms but those of the youngest children.

This book is an attempt to redress the balance, to help teachers bring back the power of imagery in work on shape and space. And although the activities are grouped into convenient sections related to mathematical ideas, there are some general principles which are complicated, but important, and which are discussed in the rest of this introduction.

Mental or practical?

The earlier activities deal with work that is entirely, or almost entirely, mental. You can give the questions to the children without the aid of any apparatus at all. But occasionally it is suggested, particularly for younger children, that some practical aid may be appropriate. The emphasis is still on the mental activity that must be carried out. The objects — the counters or triangles or walls of the classroom — are fixed, and the imagery is necessary to move them in the mind, or to construct lines or other entities between them and around them.

The later sections adopt this approach entirely: *the activity is a mental action on those objects that can be seen.* Some of the later activities are ones that children often carry out in classrooms, with no *explicit* awareness, on the part of them or their teacher, that there is mental work taking place. The activities here emphasise the fact that much mental activity is a normal part of spatial work. In *Congruent Halves* (p. 35), for example, it is possible for children to work by trial and error; but it is more likely that, in order to decide what to try in the first place, they will put in the divisions mentally, and then check them by comparing the two parts produced, rotating and reflecting in the mind where necessary. The practical alternative would be to cut the shape into two and physically turn one of the halves round or over in order to check that they fit.

The amount of practical support desirable is partly a function of the age and experience of the children. We are never sure exactly how children are able to form images. Consider the fairly simple problem of cutting the corner from a cube and deciding what shape appears (as in *Cuts*, p.21). Is it possible to 'see' what happens when you have never experienced the activity in practice? Do you calculate, in a sense, that there were three faces meeting at the original corner, and therefore the resulting shape must be a triangle? Or do you know what happens, because you have in fact performed the task physically beforehand?

Whatever it is, a great deal of practical experience is going to help. It is therefore important that younger children begin to build up this experience, so that they have a practical basis for their subsequent powers of imagery. Most of the activities in this book will be unsuitable for very young children because they do not yet have this basis.

Language

Children need to be able to talk about and describe images. There are important reasons for this.

If children cannot communicate what they are thinking, then the teacher is not sure what is going on in their minds, and cannot make the appropriate assessments that are necessary to the immediate planning of the next question, or the next activity. Mental work does not have the external signs of activity that writing or drawing does. What the children say is all that the teacher has to go on that will indicate anything about what they are doing.

Children also develop ideas by talking to each other, as well as to the teacher. Therefore it is important that discussion takes place as a vehicle for testing their own ideas, for seeing other people's points of view, and for comparing different images and strategies.

An important aid to discussion is an ever-developing mathematical language. At a simple level, it is easier to use the word 'triangle' than to have to describe a shape in terms of numbers of sides or corners. Later on it is more efficient to use a phrase like 'rotational symmetry' than to have to describe the business of, say, drawing round a shape and then turning it round to see if it still fits into its outline in a different position.

Such a technical vocabulary needs continually to be built up and shared by all the children so that discussion can take place economically and efficiently, and

so that further ideas can be developed unhampered by the need to explain again the more basic ideas.

Another handicap for very young children is that their lack of technical vocabulary makes it difficult for them to discuss their mental work.

Organisation

Mental work in geometry, and the discussion involved in it, need not require any special organisation. The activities in this book and the discussions accompanying them have been carried out either with whole classes or with groups. It depends what form of organisation you and the children are used to, and how you normally conduct discussions. Groups are obviously best, because then more children get a chance to talk. If this is not practical or desirable, then most activities can be conducted with a whole class, but a way of increasing access to the discussion is to ask the children at appropriate points to discuss what they are thinking about in small groups or in pairs; that way almost everyone at least has a chance both to talk and to listen.

Necessary or desirable?

Turning round or turning over are manifestations of two of the three basic transformations: translation, rotation and reflection. Children translate by sliding a shape along, and rotate by turning a shape round. But reflection is least easily effected practically. It needs a mirror; or else we 'cheat' by modelling the reflection by turning *over*, which is really a three-dimensional rotation or 'flip'.

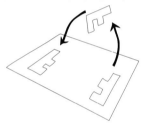

The 'flip' is an effective model at the time, but it leads to trouble when we consider transformations in three dimensions: the 'flip' is now a rotation, not a reflection. And in three dimensions, a reflection — in a plane — can be modelled *only* with a mirror.

This reflection in a mirror is useful, say, for checking that two shapes are mirror images of each other. However, it does not easily help to check symmetry. A plane of symmetry, say of a solid cube, cannot be demonstrated this way because we cannot insert a mirror *into* the cube. The *only* way of seeing it is to visualise it.

Imagery is necessary to much work in three dimensions. Children can physically construct, say, the planes of symmetry of a cube, but it is very inconvenient! I have seen the rotations in *Spins* (p.24) effected by inserting, say, a triangle into an old-fashioned egg whisk, and spinning it to produce an image of a cone, but the apparatus is not easily available, and in any case there is still some imagining to do. It is possible to construct the space diagonals of a cube using thread, but it is not easy to construct the cube in which this can be done.

However, the necessity for imagery is a red herring. The point is not so much that imagery is necessary, or even that it is convenient practically, but that it is desirable.

Algebra

There are other mathematical ideas for which mental activity is important.

The idea of algebraic generalisation can come *only* from imagery. I have asked children mentally to cut corners from a sequence of polygons, as in *Cuts* (p.21). They 'see' that a triangle becomes a hexagon, a square an octagon, a pentagon a shape with 10 sides.

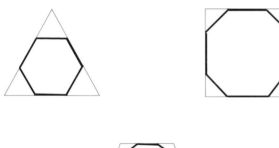

Then I ask them to imagine a shape with 100 sides, and cut the corners from that.

They know that a shape with 200 sides results. How do they know? They cannot visualise a shape with 100 sides in the sense that they can see all the sides there; but perhaps they can 'see' it in a generalised sense; they can cut every corner in a generalised sense; and they can generalise the doubling to the extent that they know that the original sides are still partly there, and that each corner has become a side. Maybe they do not go through these steps, and they merely generalise the numerical doubling in an inductive way. But whichever it is that they do, it is a mental process.

In the section on *Generalising* it is not only easier to work on the images than to make a lot of practical models, it is more effective in generalising the situations. Here is a 6-sided prism.

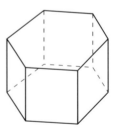

It is easy enough to count the 18 edges, but how you count them may or may not help you to generalise the situation. Now imagine a prism with 100 sides, and count the edges. If it is standing on a face, then there are 100 round the top, 100 round the bottom, and 100 vertical ones joining top to bottom. It is a small step from here to a generalisation about any number of sides: *that number* round the top, *that number* round the bottom, and *that number* up the sides. Here we have a true algebraic generalisation based on an image of the general case, and not merely induced from the pattern. Patterns can sometimes mislead.

Thinking globally

Horizontal and vertical constructions are easier to deal with than those which involve oblique lines, but the section on *Horizontal and Vertical* takes the ideas beyond the immediate field of vision and sets them, literally, in a global context. There are certain conventions about horizontal and vertical. We talk happily about a horizontal axis and a vertical axis on a graph, even though the graph paper lies on a horizontal table and both axes are therefore horizontal. We assume vertical lines on a sheet of paper are parallel, when in fact true vertical lines cannot be be parallel because they all meet at the centre of the earth. Most of

the time this is a useful convention, but it must be realised that it is only a convention. The true nature of verticality necessitates some mental imagery to take the lines to the centre of the earth, where by definition they all meet, and therefore cannot be parallel.

Having said all that, in the small part of the globe on which we live it is convenient to adopt the conventions. And it is easier to deal with many constructions if things are lined up so that the main features are horizontal and vertical. Think of a cube flat on the table, and it is easy to count edges or construct diagonals. Now think of it suspended by a corner, and try to count the corners on successive horizontal layers.

Symmetry is also a case in point. When I show a shape on an overhead screen with the axis of symmetry tilted, the viewers all tilt their heads. Children find vertical axes of symmetry much easier to identify and to work with than horizontal ones.

Because such things are easier, there is a case for presenting them this way. But there is also a case for presenting them in more difficult orientations, so that children can have experience of coping with them in these positions too.

In the same way, I have based many activities on right angles — on squares or on a square grid — and children find these much easier to deal with than activities based on other angles. So it is important that they also have experience of activities that are not based on right-angles. Polyominoes are based on squares: try polyamonds based on triangles. Nets for a cube may be old hat: try nets for tetrahedra or octahedra.

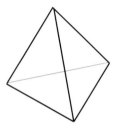 

Perhaps the one idea for which imagery is absolutely essential is the idea of infinity. (Note the famous quote from a child: "Infinity is where things happen that don't!") It is a fascinating idea for children, both spatially and numerically. The idea that numbers go on forever, or that space goes on forever, is to some extent inconceivable, yet the converse is clearly unacceptable. If there were a largest number, I could still add 1 to it. If space were bounded by a brick wall, as I once used to think as a child, there must be something on the other side of the wall. Yet we cannot experience infinity, except in the mind.

These activities and the national curriculum

The national curriculum makes no explicit mention of mental work in geometry, but it will become obvious to any teachers working with the activities in this book, if it is not obvious already, that mental work is an indispensable ingredient of shape and space.

Many of the processes involved will be recognisable in AT 1, particularly those items to do with prediction and general-isation. Consequently, the formation of rules to describe the generalisations will correspond to items in AT 3 about patterns, formulae and functions. And naturally the subject matter of the whole book is based on AT4.

One of the difficulties about attaching particular statements of attainment to the activities is that they are all suitable for a wide range of age and ability. Therefore the questions may need to be selected and modified to suit particular children.

THE
ACTIVITIES

POINTS AND LINES

These activities can be done with counters to represent the points. There is still some mental work to do in imagining how the counters are to be moved, in estimating distances, in 'filling in' a triangle, or in predicting what is going to happen.

Imagine two points

They can look fairly big, like counters. They can be black or coloured.

Move them close together. Move them further apart.
How far apart can they get?
Does it help if you imagine them further away from you?

Keep one in the centre of your vision and move the other one as far away as possible. How far away can it get?

Take three points

Put them in a straight line.

Move the middle one backwards and forwards along the line between the other two.

Put it halfway between them.

Move the middle one *off* the straight line, but so that it is still 'halfway between the other two', that is, equidistant from them. *(This will need careful explanation and some discussion.)*

Move it to different positions, keeping it the same distance, that is, equidistant, from the other two. Describe what sort of path it takes.

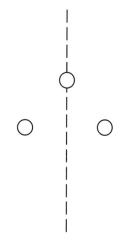

Imagine straight lines joining the points.

What sort of triangles are you making as the middle point moves?

Now let that point move to other positions. What sort of triangles can you make?

Can you make the triangle equilateral? Obtuse (one angle bigger than a right angle)?

What happens to the triangle when the third point is in line with the other two?

CONSTRUCTIONS

For some of these activities, particularly the three-dimensional ones, it may help to have a cube or other appropriate model in view. As a general rule it is probably best to try without first, in order to encourage the imagery. And in any case, there is still much imagery involved in making the subsequent constructions.

Much will depend on the age and experience of the children. Seven-year-olds, for instance, have been known to have great difficulty counting the faces, vertices and edges of a cube in their minds, even after manipulating cubes physically. And for some children the business of actually counting, say, the edges while the cube is in front of them will be a difficult mental task, because of the difficulty of organising the count. It may help to ask where the edges are, in order to encourage some sort of organisation. And if there is any problem with this, it is worth spending time on the counting of faces, vertices and edges of cubes and other shapes and leaving other activities until much later.

Imagine a cube

How many faces has it? How many corners? How many edges?

Look at one corner. How many edges go from that corner? How many faces go from that corner?

Draw a diagonal from your corner across a square face to the opposite corner. This is called a 'face diagonal'. How many face diagonals can you draw from your corner? How many face diagonals are there altogether?

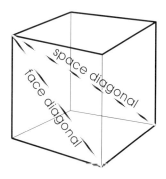

Look at your one corner again. There are three edges from it going to other corners. There are three face diagonals, also going to other corners. There is one corner left; can you see it?

Draw a line *through the cube* to that corner. This is called a 'space diagonal'. How many space diagonals are there altogether?

CONSTRUCTIONS

Start with a new cube

Put a blob of paint on one corner. How many edges have a blob at one end? How many edges have no blobs at all?

Put a blob on two corners at the ends of one edge. That edge now has two blobs. How many edges have only one blob? No blobs?

Put a blob at opposite corners of one face. How many edges now have two blobs? One blob? No blobs?

Put a blob on one corner. How many face diagonals have a blob at one end? How many space diagonals?

Put blobs at each end of an edge. How many face diagonals have a blob? How many space diagonals?

Start with a tetrahedron (triangular pyramid)

Ask the same kinds of questions as with the cube.

Imagine a square

Put an equilateral triangle on each side. Describe the shape you now have.

Imagine an equilateral triangle

Surround it with squares. Describe the new shape.

Take off the squares. Instead surround the triangle with equilateral triangles. What shape do you have now?

tetrahedron

Imagine two cubes

Join them together, face to face. Describe the shape you now have.

Join a third cube to one of the 'middle' faces — one of the faces other than the two at the ends. Describe what you have now.

Start with one cube

Put a cube on each of the faces. How many cubes are there altogether? How many faces has the new shape? How many edges? How many corners?

Start with a new cube

Put a square pyramid on each face. Describe the shape you now have.

square pyramid

Start with a tetrahedron

Put another tetrahedron on one face. Describe the shape you now have.

Start with a new tetrahedron

Put a tetrahedron on each face. Describe the shape you now have.

Join the (square) bases of two square pyramids.

Describe the shape you now have.

HORIZONTAL AND VERTICAL

The ideas of 'horizontal' and 'vertical' are normally dealt with in a fairly superficial manner. (See also the remarks in the introduction under *Thinking globally*, p.9.) Here an attempt is made to make the ideas more precise, and this requires some thought about what the ideas mean in relation to the earth.

As always, the questions about the cube may be posed with an actual cube in view.

Look at the classroom

Find some horizontal lines. Some vertical lines.
Some oblique lines, which are neither horizontal nor vertical.

Are all the horizontal lines parallel to each other? Are all the vertical lines parallel?

Imagine all the vertical lines extended downwards. Where will they end up?
Are all vertical lines parallel?

The classroom walls are vertical planes. Are they parallel?
Are any of them *almost* parallel?
The floor and the ceiling are horizontal. Are they parallel?

Are there any other planes which are horizontal or vertical?
Are there any planes which are oblique?

Imagine the earth

A flagpole is in your school playground; another flagpole is in New York. Are they both vertical? Are they parallel?

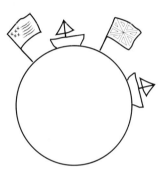

Two ships are on the equator. Their decks are horizontal. Are they parallel? When could they be parallel?

Their masts are vertical. Are they parallel? When could they be parallel?

You are in a spaceship, halfway between the earth and the moon. How do you know what is horizontal or vertical?

Imagine a rectangle

It has two sides horizontal and two sides vertical. Spin it about a vertical line. Which sides stay horizontal or vertical?

Spin it about a horizontal line. Which sides stay horizontal or vertical?

Draw in one diagonal. Can you move the rectangle so that the diagonal is vertical? So that it is horizontal?

Imagine the rectangle flat on your table, or on the floor. Which sides are vertical, or horizontal? What about the diagonals?

Imagine a cube

It is on a horizontal surface. How many edges are horizontal? How many are vertical? What about the faces? What about the diagonals (face and space)?

Imagine the cube hung up by one horizontal edge. Which faces, edges or diagonals are horizontal, vertical or oblique?

Imagine the cube hung up by one corner. Which faces, edges or diagonals are horizontal, vertical or oblique?

Imagine a tetrahedron

It is flat on the table. *(Ask the same questions.)*
Hang the tetrahedron up by an edge. What now?

What about an octahedron?

CUTS

These activities can also be carried out by looking at actual objects — plastic triangles and so on. See the section *Mental or practical?* on p.4.

As was said in the introduction, right angles are easier to deal with than other angles, so it may be better to start with the square and then go on to the triangle before dealing with the pentagon.

Note the remarks in the section on algebra in the introduction (p.8) about dealing with a polygon with 100 sides: this requires some sort of mental generalisation even if the activities leading up to it are done by looking at actual shapes.

The three-dimensional activities here are much more difficult than the two-dimensional ones, not just because they are three-dimensional, but because the complexity of the components of, say, the cube is greater. It would certainly help to have an actual cube available, but the degree of mental work left can vary according to whether the cube is merely visible, can be handled, or can be drawn on.

The discussion involved in all these activities needs to be handled with care. Two important things are that children learn to describe their images accurately, and that they continually develop a technical vocabulary for doing so. "What is it called?" is the sort of question that invites a technical term, but "How many sides has it?" asks for a confirmatory description. The second type of question, incidentally, will help to clear up any misunderstanding about terminology, like thinking that a shape with eight sides is called a hexagon.

CUTS

Imagine a triangle

How many sides has it? Are they all the same length? If not, make them all the same length.

What can you say about the corners/angles? Which way up is it? Turn it upside down. Turn it back again.

Cut a little piece off each corner, the same off each, in a symmetrical way. What shape have you got now? What is it called? How many sides has it? How many corners?

Are the sides all the same length? Are any of them the same length? Which ones? Are the corners all the same angle? What angle is it?

Can you adjust your cuts so that the sides are all the same length?

Go on cutting. What happens then?

Start with a square

Cut a little piece off each corner. How many sides has the shape now? How many corners? What is it called?

Are the sides equal? Can you make them equal?

Go on cutting. What happens then?

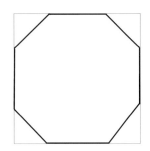

Start with a regular pentagon

Do the same thing. What happens?

Start with a regular shape which has 100 sides

Cut a little piece off each corner. How many sides now?
How many corners?

Start with a cube

How many faces has it? How many corners? How many
edges?

Cut a little piece off each corner. What have you got
now? How many faces has it? What shape are they? How
many of each shape? How many corners?

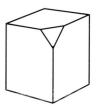

Start with a tetrahedron

How many faces, corners and edges has it? How many
edges go from each corner?

Cut a little piece off each corner . . .

SPINS

It may help to have the triangles, and so on, visible, or even for the children to be able to handle them. But this sort of help can be delayed: try it without first.

Imagine a triangle

It is an equilateral triangle, with the bottom side horizontal.

Imagine a line down the middle, an axis of symmetry. Spin the triangle around this line. Faster and faster. What sort of shape does it carve out in the air?

Turn the triangle round so that one of the sides is vertical. Spin the triangle about this side. What shape is carved out?

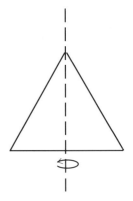

Imagine a rectangle

It has two long sides and two short sides. The long sides are vertical and the short sides are horizontal.

Spin the rectangle about one of the long sides. What shape is carved out?

Spin the rectangle about one of the short sides. What shape is produced? How is different from the other one? How is it the same?

Spin the rectangle about one of its diagonals. What shape is produced?

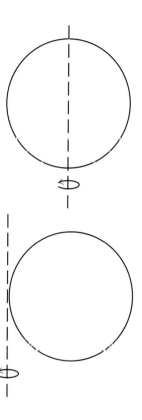

Imagine a circle

Put in a diameter. Spin the circle about its diameter. What shape is produced?

Move the diameter (a vertical one, if you like) sideways, until it touches the edge of the circle. Spin the circle about this line. What do you get?

Move the line away from the circle, so there is space in between. Spin the circle about the line. What do you get?

MOVES

These are fairly well-known activities, making 'polyominoes' from squares joined together. However, they are usually done by moving actual squares. A little more mental work is involved if the polyominoes are instead drawn on squared paper. Here it is suggested that the polyominoes are constructed entirely mentally.

One of the important things to discuss is what is meant by 'different'. When a second square is attached to the first it produces a rectangle wherever it is put, but is it the same rectangle? The children need to discuss exactly what is the same (the shape) and what is different (the orientation); and they need to decide whether the same shape in a different orientation is going to be counted as 'different', that is, is it a different polyomino?

There is also some value in identifying precisely the changes in orientation. 'Turned round' or 'turned over' may be mutually understood descriptions that are acceptable, and technical terms such as rotation and reflection may be introduced when the teacher thinks it appropriate — see the introduction starting on p.4.

Imagine a square

Imagine another square, the same size. Join them together, edge to edge.

Move the second square to another edge of the first square. Move it round to each edge in turn. Leave them joined together.

Take a third square. Join it to the pair of squares, always joining squares edge to edge. How many different places can you put it? How many different shapes can you make?

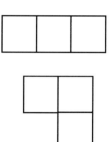

Leave the three squares joined in a row. Join a fourth square to them in different positions. How many different positions? How many different shapes?

Join the three squares to make an L-shape. Join the fourth square to them . . .

Can you cope with a fifth square?

Start with an equilateral triangle

Add further triangles in the same way as was done with the squares.

GENERALISATIONS

Making some of the prisms and pyramids is a worthwhile practical activity at any stage, perhaps some time before the mental activity suggested here, perhaps as a practical follow-up to the mental work. They can be made from one of the commercial kits of plastic polygons such as Polydron or Clixi, or by joining cocktail sticks together with small knobs of modelling clay.

The main idea of this section, however, is to produce some generalisations about the numbers of things involved, by first asking about a number so large that it can only be imagined — see the section on algebra in the introduction (p.8).

In the questions about intersections it is intended that children look for the maximum number of intersections each time. If, for instance, three lines are parallel then no intersections are produced. The sequence of maximum numbers of intersections for 2, 3, 4, 5, . . . lines is 1, 3, 6, 10, ... The differences between successive numbers of lines are 2, 3, 4, ... These differences can be explained in terms of how many lines each new line can cross. For example, the hundredth line crosses the previous 99 and therefore produces 99 more intersections.

A subsequent investigation can look at the possible ways of obtaining fewer than the maximum number of intersections.

Similar remarks apply to the questions about regions.

Prisms

Imagine a triangular prism How many faces, vertices and edges has it?

A cube is a square prism. How many faces, vertices and edges has it?

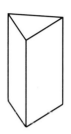

What about a pentagonal prism? A hexagonal prism? A prism with a base with 100 sides? A prism with n sides?

Pyramids

Imagine a triangular pyramid. How many faces, vertices and edges?

What about a square pyramid? A pentagonal pyramid? And so on?

Dipyramids

Join two triangular pyramids together This shape is called a triangular 'dipyramid'. How many faces, vertices and edges?

Join the bases of two square pyramids. This is a square dipyramid. How many faces, vertices and edges?

What about other dipyramids?

GENERALISATIONS

Diagonals

Imagine a triangle. How many diagonals has it?

How many diagonals has a square?

Imagine a pentagon. How many diagonals has it?

Imagine a hexagon. How many diagonals?
How many corners? How many diagonals from each corner? How many times have you counted each diagonal this way?

What about a shape with 100 sides?
How many corners? How many diagonals from each corner? . . . How many diagonals altogether?

Lines through points

Imagine two points. Draw a straight line through them.
Imagine three points, not in a straight line. Each pair of points has a straight line through them. How many lines?
What about four points? Or more? 100 points?

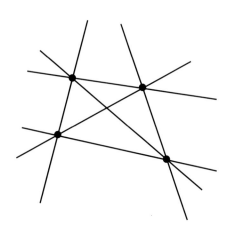

Intersections

Imagine two straight lines, intersecting. There is just one intersection.

Imagine three straight lines. How many intersections?

Four straight lines? Or more? 100 straight lines?

Regions

Imagine a sheet of paper. It is one region.

Draw a line across it, which divides it into 2 regions. If you draw a second line, how many regions can you produce?

Three lines: how many regions?

And so on.

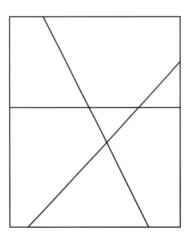

INFINITY

Imagine a square

Mark the midpoints of the sides. Join them up in order to make a smaller square.

Mark the midpoints of the sides of the new square, and join those up to make a still smaller square.

How long can you keep doing this? What happens?

Do the same thing with a triangle. Or a pentagon.

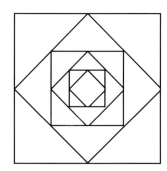

Imagine two points

Put a point halfway between them, so that you now have . . . 3 points.

Put points halfway between the new point and the old ones. How many points now?

Put points halfway between all of those. How many points now?

Carry on. How long can you keep doing this?

Imagine two squares

Put them together to make an oblong. Put two of those oblongs together to make a larger square.

Put two larger squares together to make a larger oblong.

Carry on. How far can you go?

Imagine that you are a LOGO turtle

You go forward 1, right 90, forward 2, right 90, forward 3, right 90, and so on, increasing your forward movement by 1 each time.

What happens? Where do you end up? (Assume you are not limited by the screen!)

Imagine a right-angled isosceles triangle

Cut it in half, to make two smaller similar triangles.

Cut one of those in half. Cut one of those in half. How far can you go on?

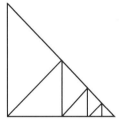

Imagine a circle

It grows, bigger and bigger. How big can you make it?

Now start again with a small circle. Make it bigger, but stay on the circumference, so that the centre moves away.

Keep making it bigger, with the centre getting further away and you staying at the circumference.

What happens to the centre? What happens to the circumference?

Here are two infinitely long lines

(Two metre rulers could be used. The children will have to imagine that they are infinitely long.)

The bottom line is fixed. The top line is pivoted at the point A, and it intersects the bottom line at the point B.

Rotate the top line slightly, so that the point of intersection B moves to the left.

Rotate a little more, so that B goes further to the left.

Carry on, making B go further and further to the left. How far to the left can you make it go?

What happens when the lines are almost parallel? When they *are* parallel? When the line continues to rotate?

Can you reach the end of the universe?

If not, why not?

If you think you can, what happens if you go a little further?

CONGRUENT HALVES

Each of the shapes must be separated into two parts which are exactly the same. Some can be done in more than one way.

The pages will need to be photocopied so that the children can work on the actual shapes. Pencil and rubber would allow for alterations. Alternatively the shapes could be copied, by the children, onto geoboards (a minimum of 6 x 6 pins is necessary) and the divisions made with rubber bands, which allows for easy alterations.

The geoboard copies could be made from the photocopied sheets, or from an overhead transparency. The transparency could also be useful for a class discussion of solutions.

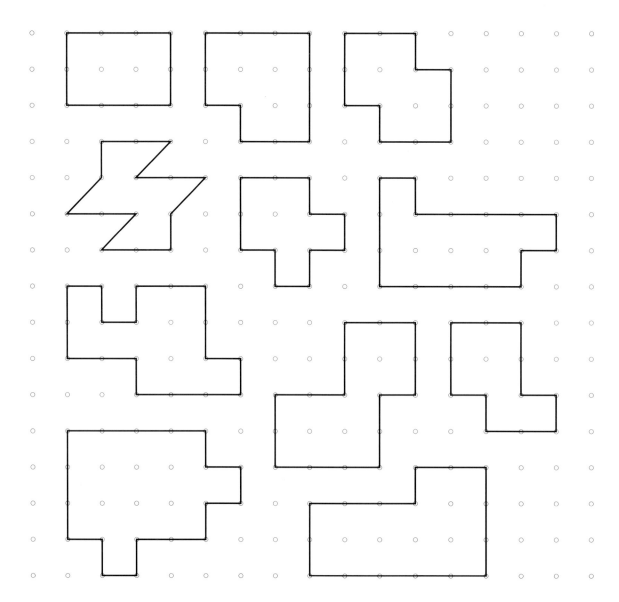

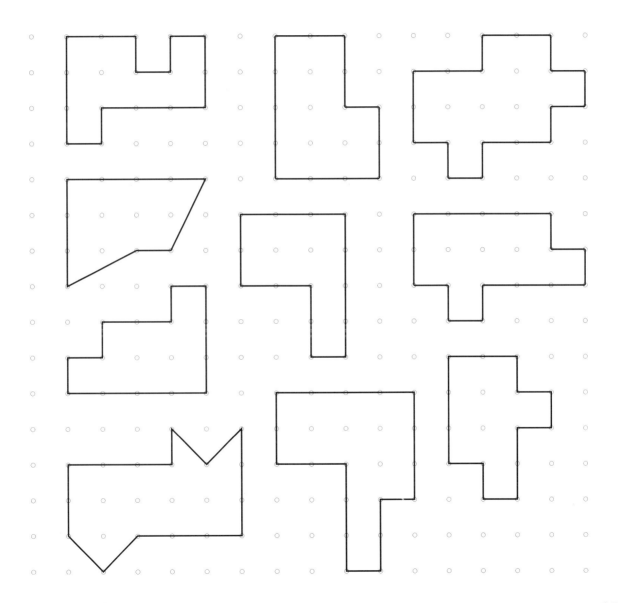

CONGRUENT HALVES

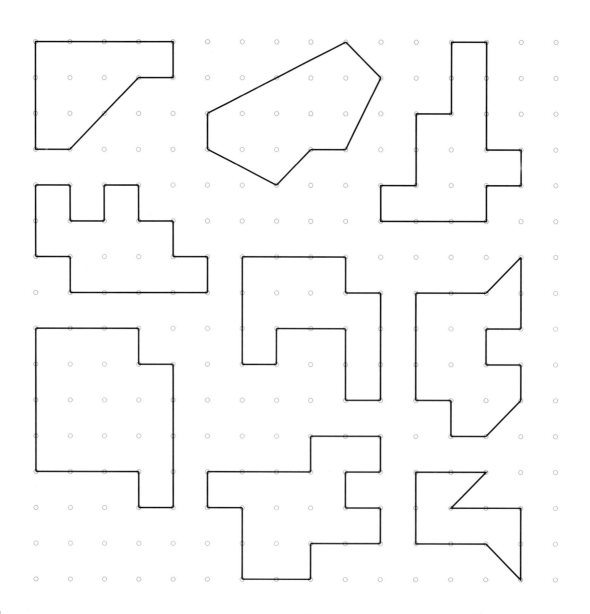

POINTS AND SHAPES

Counters can be used to represent the points.

Here are the corners of a triangle.
Can you 'see' the triangle?

Can you see a triangle here?

How many triangles can you see here?
How many quadrilaterals can you see?

40

How many quadrilaterals can you see now?

How many triangles here?

Or here?

Investigate how many triangles can be made with different numbers of counters.

What about quadrilaterals?

NETS

These can be presented on an OHP or blackboard, or the pages from this book can be reproduced.

If this is presented as a class or group activity it can be interesting to ask which edges or corners come together when the nets are folded up to make the cubes, and possibly helpful to the children. In describing this, children often think of one square as the base of the cube and in their minds fold up the sides and then the top. Such strategies are worth sharing.

Which of these nets will fold up to make a cube? If any will not, then say why not.

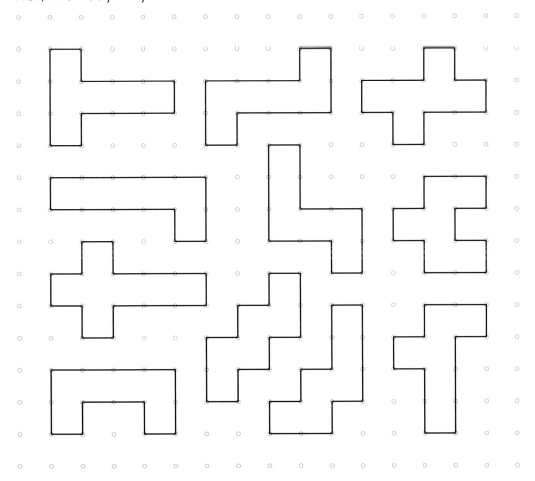

How many different nets for a cube are there?

SYMMETRY

Put in the lines or centres of symmetry in these shapes.

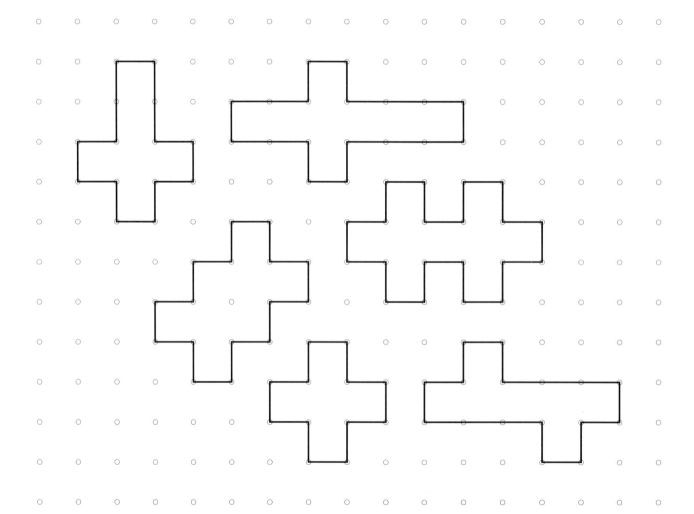

And these.

SYMMETRY

Add one square to each of these shapes to make it symmetrical.

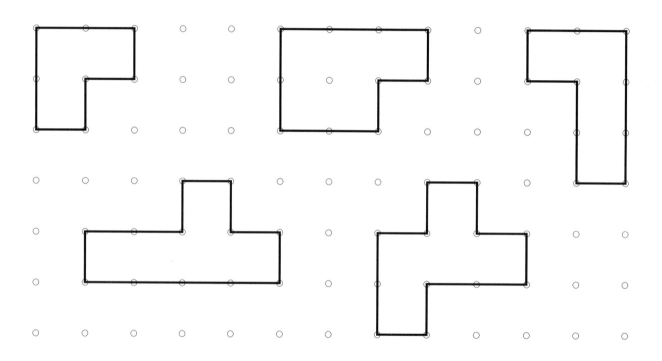

How many ways are there of doing it for each one?

Add one small triangle to each of these shapes to make it symmetrical.

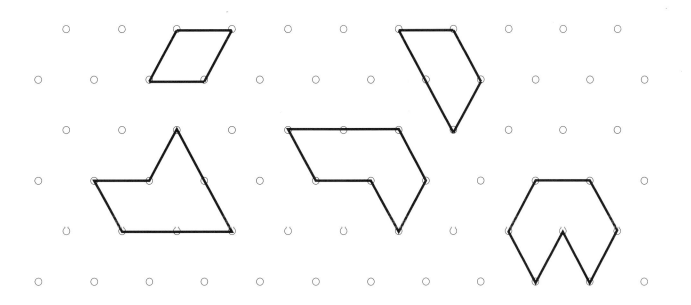

How many possibilities are there for each?

PAPER AND SCISSORS

Although these are very practical activities, the important thing is the prediction of what will happen. In the initial activities with one fold, which incidentally are very suitable for Year 1, the children must think in advance about the shape they want to produce.

In the activities with two or more folds the children should be asked to think and talk about what will happen before they start to unfold their paper, giving their reasons, and describing the shapes that will appear.

When asked, say, to describe exactly how to cut the paper with two folds in order to produce a square, children are often vague in their instructions as to where to put the scissors: "move them up a bit", "turn them round". By refusing to understand what is meant the teacher can encourage more precise descriptions in terms of distance or angle: "how far up?", "how much round?"

Fold a sheet of paper in half

Cut something out of the fold, so that when you unfold
the paper you will have a square hole.

Try again, but this time end up with a triangular hole. What
about a circle? A shape with five sides?

Fold a sheet of paper in half, and then in half again

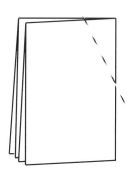

Cut off the corner, like this.

What shape will the hole be when
you unfold the paper? Think about
it before you unfold.

How can you cut to make
a square? A rhombus?

Fold a sheet of paper in half, in half again, and then like this, to make an angle of 45°

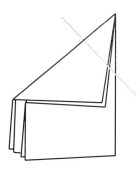

Cut off the corner.

What shape will you have when
you unfold?

What different shapes can you get,
always with one straight cut?

How can you fold a piece of paper so that you can make
an 8-pointed star with one straight cut?

LESSON

(Extracts from a lesson with a group of ten-year-old children. See Paper and Scissors, page 48.)

DF Can you fold your paper in half, and fold it in half again, like that?

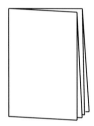

Now I've got that corner that was originally in the centre of the paper. Watch me carefully, because I'm going to cut that bit off.

What shape am I going to get when I open that up?

Ellie A triangle.

Gavin A square.

Daniel A diamond.

Jago A triangle or a diamond.

Ella I think you'll get a diamond.

DF What's a diamond?

Daniel A pentagon.

DF What's a pentagon?

Daniel A five-pointed shape.

DF What's a diamond?

Daniel More like that shape (*indicating with fingers*).

DF Tell me about it.

Lily Four sides. A kind of wonky square.

Gavin A square would be like that (*holding his paper with sides horizontal and vertical*). A diamond would be like that (*holding it with sides oblique*).

Margaret A diamond's got four corners . . . and that part's sort of . . . longer.

DF How many sides has a diamond got?

Margaret Four.

DF And can you tell me anything about the sides?

Lily Same size.

Jago   All the same size.

DF   So if I open this up . . .

. . . that's what you call a diamond, is it?

Daniel   A rhombus.

DF   Yes, that's the proper name for it, a rhombus.

(The children cut a similar rhombus from their own paper.)

The children have had time to allow their mental imagery to work, and are able to offer various suggestions. The queries from the teacher gradually probe for information about what the terms mean: "what's a diamond?" "how many sides . . ?" "tell me about it". 'Diamond' is not a technical mathematical word, and when Daniel provides the term 'rhombus' this is accepted.

DF   OK. I'm going to do the same thing again, fold this in half . . . and in half again. *(This time the cut is made at 45° to the edges of the paper.)*
What do you think that will be when I open it up?

Lily and Ella   A square.

DF   Why is it going to be a square?

Daniel   How many times did you fold it?

DF   The same as before.

Jago   It'll be the same sort of shape.

Ellie   Before, it was stretched, so it couldn't have been a square. But this time it's sort of, like, a corner of a square more.

DF   So before it was stretched?

Ellie   A stretched square. But this time you've cut it differently, so . . .

DF   Can you say how I cut it this time?

Ellie   You cut it sort of . . . smaller . . . it's like the corner of a square. Before it wasn't.

DF   This is what I cut before.

So that's not the corner of a square?

Ellie   No.

DF   How do you recognise the corner of a square?

Daniel   Well, it'd be a parallelogram.

DF   A parallelogram?

Daniel   No, I mean the other one.

DF   What, this one?

This is the one I did first time.

Daniel   No, that's not a parallelogram.

DF   Is this the corner of a square?

Jago   No. Not a corner. Sort of . . . part of it. It could be, but it isn't. If you open it up, you'll have a . . . sort of like . . . like this but a bit fatter, and not as long.

Ellie   You could turn it into a square . . . You can get that shape . . . It doesn't have to be. If you turn it round it'd be a square.

Most of the children recognise that it will be a square, but have difficulty in explaining why. They give circular arguments, "it's the corner of a square", or contrasts with the previous shape, "before, it was stretched."

(The square is opened up, and the teacher now asks how he has to cut the folded right angle in order to obtain a square.)

Jago   Make it a half right angle, across there.

Lily   It's got to be equal.

Ellie   So that it's equal.

DF   So what's equal?

Ellie   Cut at a slant.

DF   Is that at a slant?

Several   Yes.

Lily   So those two sides are the same.

DF   Which two sides?

Lily   That and that (*pointing*).

DF   If I shut my eyes, can you tell me which two sides?

Lily   The top, and along to the right.

DF   Is that it?

All   Yes.

The children's instructions are imprecise: "equal", without saying what must be equal; "at a slant", without specifying the

*angle; "those two sides the same",
without saying which until pressed.*

*(The teacher folds up another sheet of
paper as before, then, with discussion
about it, into a half right angle, and
prepares to cut that.)*

DF   What shape am I going to get when I
open that up?

Lily   A rhombus.

Ellie   I can't remember what it's called.

DF   Can you tell me how many sides
it's got?

*A crucial question, before the children get
unnecessarily involved in terminology.*

Lily and Ella   A rhombus.

Ellie   A shape with five sides.

DF   Is that a rhombus?

?   No.

Ellie   Five sides.

Ella   I think six sides.

Ellie   Five or six.

Lily   Eight.

Ella   It'll be eight, wouldn't it?

DF   Why?

*The teacher eventually has to ask why, to
discourage the guesses, and to
encourage the intuitions to be analysed
more carefully.*

Ella   Because if you cut it like that, before
it'll be four sides, and if you fold it like that,
it's going to be eight sides.

Lily   Probably. If you count these bits of
paper here, there's eight of them.

DF   *(Cuts.)* How many bits of paper?

Daniel   Four.

*Both Ella and Lily are reasoning. But
Daniel's wrong answer needs checking.*

DF   If we think how we folded it up . . .
(taking another sheet of paper) . . . I had
one thickness of paper to start with,
didn't I? I folded it in half; how many
thicknesses now?

All   Two.

DF   If I fold it in half again, how many
thicknesses?

Daniel   Three.

Others   Four.

DF   Three or four?

Several   Four.

Daniel   Three.

DF   How many thicknesses were there?

All   Two.

(*The teacher folds again.*)

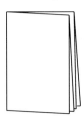

All   Four.

DF   Everybody agree it's four?

All   (*Including Daniel!*) Yes.

DF   And if I fold it again, how many thicknesses of paper?

All   Eight.

DF   So I cut this piece off.

When I unfold that, how many sides is the shape going to have?

All   Eight.

DF   And what do we call an eight-sided shape?

Daniel   A octagon.

DF   An octagon. Right.

(*The octagon is opened up, and then the children cut their own. Now the teacher prepares to cut another half right angle, but at a different angle.*)

DF   So that's going to be an octagon when I open it up?

Lily   Yes.

Daniel   No.

(*The teacher waits, then unfolds once.*)

Jago It's not the same.

(*The teacher unfolds once more.*)

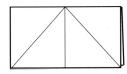

DF What's this going to be?

(*This time it is a square. They discuss inconclusively the paradox that in spite of the eight thickness of paper, a square is produced.*)

DF Can you tell me how to cut so I get a square?

Jago Going down that way, and that way. Both exactly the same.

This is one of those frequent occasions when the teacher knows exactly what the child means, but because it has been put imprecisely he can pretend not to understand.

DF What does that mean? If I cut . . .

Jago No.

Lily Cut straight.

DF That is a straight cut, isn't it?

Jago No. Have both of the sides straight.

(*This discussion goes on for some time, with the children continuing to use "straight", "slant", "move the scissors", "turn the scissors" and "straight across".*)

DF When you say straight across, what do you mean?

Ella It isn't actually straight, but . . .

Ellie A bit more straight.

Daniel Like that.

DF Like what?

Ellie Horizontal.

DF What's got to be horizontal?

Ellie The scissors.

DF What about the paper?

Ella It's got to be in between the scissors! (*Laughter.*)

DF Yes! I've got the scissors horizontal. How do you want me to have the paper?

Ella Straight, like that. That bit pointing upwards.

DF Which bit? I thought it *was* pointing upwards.

Ellie Keep that side there going diagonally.

DF What do you mean, diagonally?

Ellie Going diagonally . . . Straight. This side straight, and that side's diagonal.

DF They're both straight sides, aren't they?

Ellie   No, I mean straight down. That side straight down.

DF   Oh, straight down. What do we call something that's straight down?

Ellie   Um . . .

Ella   Vertical.

DF   So, I want the scissors horizontal, and what do I want vertical?

Ellie   That . . . the right side. That'll be it. Yes.

DF   If I have one side of the paper vertical and the scissors horizontal, what's the angle between them?

Ellie   A right angle.

One feature here is the ambiguity of everyday words like 'straight', 'slanting' and 'diagonally', compared to the precision of mathematical words like 'horizontal' and 'vertical'.

But another important feature is the difference between the 'local' relationship between the scissors and the edge of the paper (the right angle), and the more distant relationships like horizontal and vertical, where paper or scissors are related to the external world. These children appear to find it easier to visualise the latter.

This, incidentally, is in contrast to the behaviour of younger children (usually up to the age of 4 or 5 years), who concentrate more on the immediate relationships. This phenomenon manifests itself for example in the drawing of chimneys at right angles to sloping roofs instead of at right angles to the ground, or in Piaget's experiment where model trees and posts are placed at right angles to the sides of a clay 'mountain'. These phenomena are the results of mental activity, since the children are creating what they imagine, rather than what they actually see.

The older children described in the lesson are also in a constant state of mental activity, as they involve themselves with a struggle with verbal descriptions of their images and their predictions.